SCIENCE AT WORK

MAGNETISM

AT

WORK

REBECCA FELIX

Consulting Editor, Diane Craig, M.A./Reading Specialist

Sandcastle

An Imprint of Abdo Publishing
abdopublishing.com

Published by Abdo Publishing, a division of ABDO, PO Box 398166, Minneapolis, Minnesota 55439. Copyright © 2017 by Abdo Consulting Group, Inc. International copyrights reserved in all countries. No part of this book may be reproduced in any form without written permission from the publisher. SandCastle™ is a trademark and logo of Abdo Publishing.

Printed in the United States of America, North Mankato, Minnesota

062016
092016

Design: Mighty Media, Inc.
Content Developer: Nancy Tuminelly
Production: Mighty Media, Inc.
Editor: Liz Salzmann
Photo Credits: Shutterstock, Wellcome Trust

Library of Congress Cataloging-in-Publication Data

Names: Felix, Rebecca, 1984- author.
Title: Magnetism at work / Rebecca Felix ; consulting editor, Diane Craig,
   M.A./reading specialist.
Description: Minneapolis, Minnesota : Abdo Publishing, [2017] | Series:
   Science at work
Identifiers: LCCN 2015050533 (print) | LCCN 2016000128 (ebook) | ISBN
   9781680781427 (print) | ISBN 9781680775853 (ebook)
Subjects: LCSH: Magnetism--Juvenile fiction.
Classification: LCC QC757.5 .F45 2017 (print) | LCC QC757.5 (ebook) | DDC
   538--dc23
LC record available at http://lccn.loc.gov/2015050533

SandCastle™ Level: Fluent

SandCastle™ books are created by a team of professional educators, reading specialists, and content developers around five essential components—phonemic awareness, phonics, vocabulary, text comprehension, and fluency—to assist young readers as they develop reading skills and strategies and increase their general knowledge. All books are written, reviewed, and leveled for guided reading, early reading intervention, and Accelerated Reader™ programs for use in shared, guided, and independent reading and writing activities to support a balanced approach to literacy instruction. The SandCastle™ series has four levels that correspond to early literacy development. The levels are provided to help teachers and parents select appropriate books for young readers.

EMERGING · BEGINNING · TRANSITIONAL · FLUENT

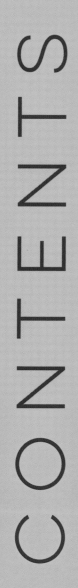

CONTENTS

ABOUT MAGNETISM

Do you use a refrigerator?
Have you ever used a **compass**?

Then you have
used magnetism!

Magnetism is a force. It is found
in some metals.

They pull toward
each other.

Scientists study
magnetism.

Michael Faraday was one.
He experimented with magnets.

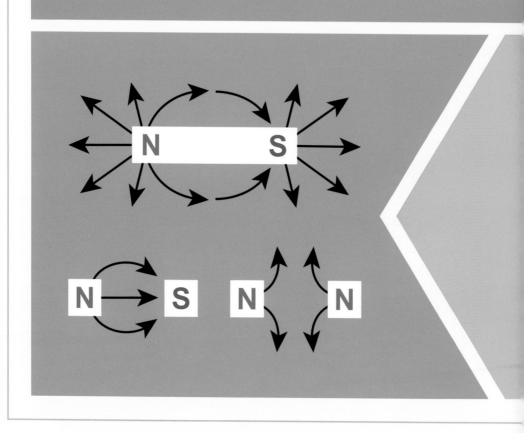

Magnets have two poles.
North and south poles **attract**.

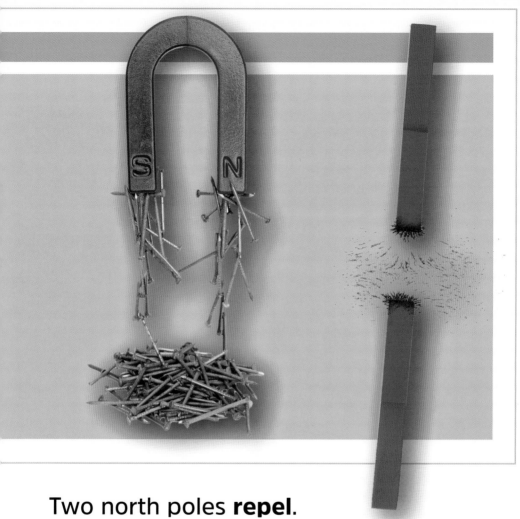

Two north poles **repel**.
So do two south poles.

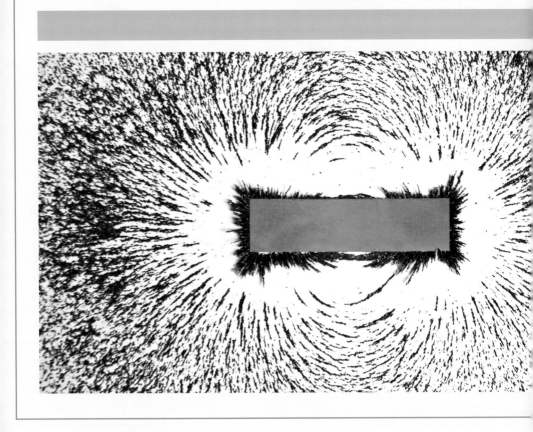

A magnet's force surrounds it.
This area is called a magnetic field.

Metals move
toward it.

13

People can create
magnetic fields.

They make magnets too.
They use them in many ways.

Magnets are used in TVs.
They are used in **motors** too.

Some **cranes** use huge magnets
to lift heavy objects.

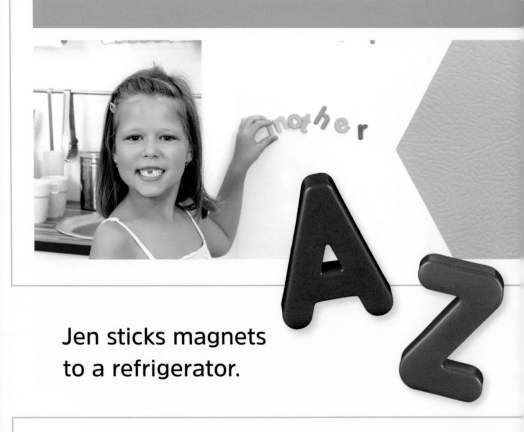

Jen sticks magnets
to a refrigerator.

The refrigerator also has a magnet
in its door. It holds the door shut.

Mason has a **compass**.
Its needle has a magnet.

It moves toward metals in the earth.

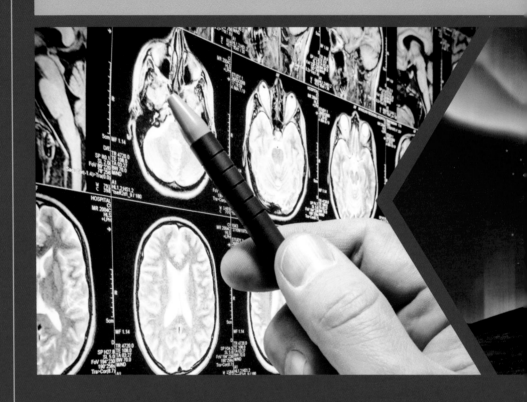

THINK ABOUT IT

Look around you! Where else is magnetism at work? How do you use it?

GLOSSARY

attract – to cause something to move closer.

compass – an instrument used to find directions. A compass has a magnet that always points north.

crane – a machine with cables and a long arm that is used for moving heavy items.

motor – a machine that creates motion or power.

repel – to drive or push back or away.